Red Bay, Labrador

World Whaling Capital A.D. 1550-1600

James A. Tuck
Robert Grenier

Published by Atlantic Archaeology Ltd.
St. John's, Newfoundland, Canada

ISBN 0-929048-00-8

Cover illustration by Richard Schlecht courtesy of National Geographic Society

Printed in Canada

CONTENTS

Red Bay, Labrador

Red Bay, today a small fishing community on the southern coast of Labrador, was at one time the largest whaling port in the world. From the 1530s until the early 1600s ships and men from the Basque Provinces of south-western France and northeastern Spain came here each year to hunt the large right and bowhead whales which migrated annually through the narrow Strait of Belle Isle. The Strait was known to sixteenth century Basques as 'The Grand Bay,' what is now Newfoundland and Labrador as 'The Province of Terranova,' and Red Bay as 'Butus' or 'Buteres.'

Red Bay in early June. Pack ice is still visible in the harbour and in the Strait of Belle Isle beyond. Saddle Island shows clearly the reason for its name. Whaling stations are located on Saddle Island, on the small island to the right and on the mainland to the left.

Denis Pagé photo courtesy of Parks Canada.

Although the existence of Basque whaling in the area has been known to historians and other scholars for many years, it was not until the late 1970s that the truly industrial proportions of sixteenth century whaling in southern Labrador were recognized.

More than a dozen whaling ports, scattered along the coast from Middle Bay in Quebec to Cape Charles, Labrador, have been reported by archaeologists and historians. Most of these are small harbours to which between one and three ships might have sailed each summer. Archaeological discoveries around the harbour at Red Bay, on the other hand, have revealed evidence which suggests that each year during the latter half of the sixteenth century as many as fifteen ships, and perhaps even more during some years, made the voyage from Europe to this port to hunt whales. This booklet relates the story of the discoveries made along the coast since 1977, with particular reference to the excavations at Red Bay, which can justifiably claim the title of having been the 'World Whaling Capital' between about A.D. 1550 and A.D. 1600.

Red Bay as it might have appeared 400 years ago. A northeast wind blows smoke from the rendering stations across Saddle Island and ships riding at anchor in the harbour. The ships, which did not take part in the whale hunt, have their top rigging removed to prevent damage from the wind. The shore stations on Saddle Island at the extreme right and left were excavated by archaeologists between 1979 and 1988.

Painting by Richard Schlecht courtesy of National Geographic Society.

Whales: Vanished Species

A humpback whale leaps from the Strait of Belle Isle. Humpbacks are today among the most common whales in Labrador waters.

Denis Pagé photo courtesy of Parks Canada.

Visitors to Red Bay and the Strait of Belle Isle area today are very likely to observe whales in near-shore waters. The most common species are the relatively small minke whale and the much larger, and more acrobatic, humpback. Other species which might be sighted are fin whales, sei whales, killer whales and giant blue whales. Notably absent from this list are the two species of whales which were hunted by the Basques—right whales and bowhead whales.

The right whale was named for its behavioural and physical characteristics. It is slow-moving, easy to approach, floats when killed, and yields great quantities of oil and baleen—in short it was the 'right' whale to hunt. The bow-head whale possesses these same characteristics, including an even greater yield of baleen—the plates which are suspended from the whales' upper jaws through which they strain seawater to obtain food. This substance, often called

Tail flukes of a right whale emerge as it begins its dive. Once a common sight in Labrador waters, only a few right whales remain in the North Atlantic.

Photo courtesy of Scott D. Kraus, New England Aquarium.

'whalebone,' was used by former generations for a variety of products that required both strength and elasticity, for example umbrella ribs and corset stays.

Bowhead whales are today confined to arctic waters; a small population of right whales now summers in the Gulf of Maine and Bay of Fundy. Centuries ago, however, both right and bowhead whales migrated regularly through the Strait of Belle Isle. Dr. Stephen Cumbaa, of Canada's National Museum of Natural Sciences, has studied the whale bones from the Red Bay sites and made the following suggestions about the whales' migration patterns. Right whales migrated northward during the spring and early summer and arrived in the Strait of Belle Isle in early July where they remained until the onset of cooler conditions signalled the time to return south, probably in late August or early September. Bowhead whales are an arctic species, which travelled southward ahead of the pack ice and probably arrived in the Strait of Belle Isle in late October, where they remained for several months.

The Basque whalers, therefore, had two distinct seasons during which whales were available. This fact is reflected in documents written in the sixteenth century: they often state that if the crew of a whaling ship was not successful in obtaining a full cargo of oil during the early season they were to remain in Terranova for the late season.

Unless there were some bowhead whales which remained in the area throughout the winter, and therefore were available to the whalers early the following summer, it seems as if the late season, during which bowhead whales were normally hunted, was the more important of the two seasons—more than one-half of the identifiable whale bones from the Red Bay sites are those of bowheads.

Cloud of vapour marks the exhale of a right whale as it surfaces to breathe.

Photo courtesy of Scott D. Kraus, New England Aquarium.

It is not known how many right and bowhead whales the Basques killed during the eighty or so years that they plied their trade along the north shore of the Strait of Belle Isle. It is certain, however, that the numbers were very large and most estimates range into many thousands. For slow breeding species such as whales, a kill of even ten thousand would have put a severe strain on a population. It is very likely that the Basque predation on right and bowhead whales started a population decline from which the whale stocks have never recovered. Although a number of factors have been held to account for the cessation of Basque whaling in the Strait of Belle Isle in the early 1600s, it seems likely that overhunting was the primary cause for the demise of the whale fishery.

Rare photograph of a bowhead whale and calf was taken from the air above arctic waters to which these animals are now confined.

Photo courtesy of David J. Rugh, National Marine Mammal Laboratory, NMFS, NOAA, Seattle.

Whalers

Seals from Basque towns, some dating from the Middle Ages, illustrate the importance of whaling to the economy of the Basque country. Shown here are the seals of Bairritz and Motrico, the former dating to the year 1351.

It is not certain when, and where, in the world whaling began. In all likelihood it was developed in several places by various peoples. The Basques of the Bay of Biscay in Spain and France, however, may be credited with the development of the type of whaling familiar to readers of *Moby Dick* and other eighteenth and nineteenth century accounts of British and American whaling.

As early as the eleventh or twelfth century Basques are reported as taking large Biscayan right whales, close relatives of the whales they later hunted in the Strait of Belle Isle. Initially the whales were hunted for their meat, which was salted and shipped to many parts of Europe. Eventually, however, the oil which could be rendered from the whales' thick layer of blubber became a far more important product. In time it came to be used for lighting, soap-making, in the cloth and leather industries and for a variety of other purposes.

A mountain village in the Basque country. Men from remote villages such as this travelled each year to Terranova in the late sixteenth century to hunt whales.

Robert Grenier photo courtesy of Parks Canada.

The men who manned the whaleboats which set out from the shores of the Bay of Biscay are today known to us as Basques, and to themselves as *Euskaldunak*. The Basque people inhabit the Pyrenees Mountains and the shores of the Bay of Biscay in southwestern France and norththeastern Spain. The Basque language appears to be unrelated to any known European or other language, and even some biological characteristics, such as blood types, also set the Basques apart from their neighbours.

In North America Basques are often thought of as the shepherds of the American west, since many of them immigrated to the United States in the nineteenth century to find employment on sheep ranches in Nevada and neighbouring states.

In fact the Basque people have traditionally been as much involved with ship building, fishing and whaling as they have been with the raising of sheep and the manufacture of woolen cloth. Many of the ships which made the annual voyages from Spain to the

Caribbean and back throughout the sixteenth and seventeenth centuries were built of Basque oak in ports such as San Sebastián or Santander; their fastenings, anchors and armaments were of iron smelted and forged in the Basque country. Many of the early French voyagers and fisherman who visited Atlantic Canada were in fact Basques who set sail from St. Jean-de-Luz or Bayonne.

Basque fishermen were among the first Europeans to visit Atlantic Canada and to exploit the vast stocks of codfish discovered there. Jacques Cartier met Basques familiar with the Strait of Belle Isle during his voyage in the 1540s. The cod fishermen must have been quick to notice the large schools of whales which passed through the Strait of Belle Isle, some of which were identical to those they had hunted in the Bay of Biscay. At the same time as the New World was being discovered, the stocks of whales in the Bay of Biscay were beginning to dwindle; the fortunate discovery of whales in the Strait of Belle Isle may have provided a much-needed alternative to the decreasing whale fishery in Europe.

Basque shipyards such as this produced the ships and boats used in the sixteenth century whale hunt.

Robert Grenier photo courtesy of Parks Canada.

By the mid-sixteenth century Basque whalers were well established at a number of ports on the north shore of the Strait of Belle Isle—known to the Basques as the 'Grand Bay.' Here they hunted whales for more than a half century when stocks once again began to dwindle and whaling became unprofitable. Events in Europe—the defeat of Philip's Great Armada, which included many Basque ships and men, and inflation, which made it difficult to put together capital for whaling voyages—appear to have made it impossible for the Basques to maintain the virtual monopoly on whaling which they had held for centuries. Nonetheless, Basque involvement in whaling did not cease.

Indeed, it was the Basques who taught the techniques of whaling to the nations who were to succeed them as the world's great producers of whale oil and baleen. Following the Terranova whaling, the focus shifted to the arctic waters of the Greenland and Barents Seas at Jan Mayen Island and at Spitsbergen. While a few Basque ships appear to have visited these locations, the principal involvement of Basques in seventeenth century whaling was as paid members of whaling crews who taught their age-old skills to the Dutch and English. It seems quite clear that these basic skills—the hunting of whales and rendering of their oil—can then be traced forward in time to the whaling of the eighteenth and nineteenth centuries and, in modified forms, to the modern whaling era.

Harpooners enjoyed both high wages and prestige as a result of their skills.

Painting by Richard Schlecht courtesy of National Geographic Society.

Whaling

The techniques employed by the sixteenth century whalers at Red Bay probably differed very little from those employed for generations along the coasts of the Basque country. When the whalers arrived at Red Bay in early summer they immediately set about constructing or repairing shore stations where the whales would be flensed and the blubber rendered into oil. The whaleships, called galleons or *naos*, were moored safely in the harbour and the top rigging removed to make them less susceptible to storms. Here they rode at anchor and served as floating warehouses until it was time for the return voyage at the close of the season. Even while this work was under way lookouts were posted on strategic hills along the coast to scan the horizon for the first approach of whales.

A right whale, struck by the lance, makes its last dive. The harpooner in the bow of the boat holds a second lance in case the first missed its mark.

Painting by Richard Schlecht courtesy of National Geographic Society.

A woodcut from an early description of whaling shows most of the elements of shore whaling. Whales are towed to shore for flensing and rendering; the oil is stored in wooden casks. What it fails to represent is the truly industrial scale of sixteenth century whaling in Labrador.

When a whale was sighted crews immediately set off in pursuit in small boats called *chalupas*, or in English 'shallops.' These boats averaged about eight metres (26 feet) in length and were manned by five or six oarsmen and a steersman. The whale was harpooned with a barbed iron harpoon attached to a sturdy oak shaft. A 'drogue,' or drag, attached by a rope to the harpoon was thrown overboard from the whaleboat. Towing the drogue slowed and tired the whale and as it surfaced to breathe additional harpoons and drogues were made fast. None of the whaleboats recovered by Parks Canada Marine Archaeology Unit divers from Red Bay Harbour has provision for making the harpoon line fast to the boat itself, so it is not certain whether the famous 'Nantucket sleighride' of more recent centuries was a part of sixteenth century whaling.

Eventually the whale became exhausted and could once again be approached by the small boats. When this occurred a razor-sharp lance was thrust into the whale's vital organs to kill it as quickly as possible. This was the most dangerous part of the hunt and no doubt many boats were capsized by thrashing whales as they attempted to escape the entangling lines and approaching boats. Up to this point sixteenth century whale hunting was remarkably like that which persisted until the advent of the harpoon gun in the late nineteenth century.

Once the whale was dead it was not towed to the waiting ship for processing, as in later deep sea whaling, but was laboriously towed by rowing to one of the stations constructed along sheltered shores of Red Bay Harbour. Favourable winds occasionally allowed the use of sail in this process but the strong tides and often adverse winds must have made this an ordeal, even when several boats combined to tow the whale's carcass.

At the shore station the whale's flippers and flukes (tail) were removed to allow the carcass to be rotated in the shallow water where long strips of skin and blubber were removed. The blubber was

A harpoon, probably broken during the hunt, was recovered from Saddle Island. Originally about a metre long, the end opposite the point was equipped with a ferrule into which a stout oak shaft was fitted.

then cut into smaller chunks and finally 'minced' into small pieces, often using a flipper or fluke as a cutting block.

The minced blubber was placed in one of a number of large, shallow copper cauldrons arranged atop a series of stone fire-boxes which comprised the 'tryworks' where the actual rendering took place. A fire was kindled beneath each cauldron and a slow heat rendered the blubber into oil. The oil was then ladelled from the cauldrons into tubs partly filled with water; the lighter oil floated and the impurities settled. The oil, by this time both cooled and purified, was finally ladelled into casks and stowed carefully aboard the galleons for the return trip to Europe. The bits of skin and other tissue which would yield no more oil were removed from the cauldrons and used to fuel the fires beneath. No wood or other fuel was necessary; in a sense the whales boiled themselves down.

Rediscovering Butus

Bones of right and bowhead whales still litter the shore of Red Bay Harbour giving testimony to the presence of Basque whalers 400 years ago.

For almost four centuries the Basque settlement at Butus lay forgotten and overgrown by the meagre vegetation of southern Labrador. Fragments of the red roofing tiles used by the Basques to cover some of the buildings were continually exposed on the beaches around the harbour where they were noticed by generations of Red Bay children. Little thought was given the 'red rocks,' except for their usefulness in writing on rocks and making paint; few people living in Red Bay today did not experiment with them.

In the mid-1970's the red rocks so common on the beaches of Red Bay began to take on a new meaning. Historical geographer Selma Barkham, working in long forgotten archives in the Basque region of Spain, began to discover thousands of documents involving Basque fishing and whaling in Terranova. Of particular interest were documents which referred to whaling in the Grand Bay and especially at the port called 'Butus' or 'Buteres.' Using these documents, and a set of sailing directions published in 1576 by a Basque navigator, Martin de Hoyarsabal, Mrs. Barkham was able to establish the locations of several of the sixteenth century whaling ports, including Butus.

In 1977 Mrs. Barkham accompanied a field party from Memorial University of Newfoundland and the Royal Ontario Museum to southern Labrador to investigate her discoveries on the ground. Verification of the locations of several Basque ports followed quickly. Particularly impressive was the number of red roofing tile fragments which littered the beaches and gardens in

Four hundred years ago the shorefront of Saddle Island, here seen from the mainland, was lined with smoking tryworks producing valuable whale oil. On the small level terraces above the beach were the tile-roofed buildings where coopers worked and lived.

the community of Red Bay and on Saddle Island, which shelters Red Bay Harbour from the Strait of Belle Isle.

Also on Saddle Island there were found the remains of low crumbling stone walls still bearing thick deposits of a hard black substance that could only have been burned whale fat. Test excavations around the stone walls revealed thick layers of broken red tile, clearly the collapsed roofs of sixteenth century structures. Other test excavations produced small sherds of pottery also typical of the sixteenth century.

Since 1977 field crews from Memorial University of Newfoundland have explored the remains of these early European whaling establishments, which have proved to be far more extensive than originally anticipated.

In 1978 marine archaeologists from Parks Canada conducted a survey in Red Bay Harbour, again prompted by Mrs. Barkham. In this case the archaeologists sought the remains of the *San Juan*, a Basque

Oak timbers protruding through the silt of the harbour floor gave Parks Canada divers their first view of a Basque whaleship from the sixteenth century.

Robert Grenier photo courtesy of Parks Canada.

whaler lost in the late fall or early winter of 1565 just as she was preparing to leave, with a full cargo of whale oil, for the return voyage to Spain. For the next eight years Parks Canada divers worked in the cold waters of Red Bay Harbour where they discovered the remains of two more ships and a number of small boats and excavated completely the hull of the ship that is probably the *San Juan*.

The documents which led to the discovery of the whaling port of Butus and the galleon *San Juan* are of a type called 'notarial' documents; that is, they were attested to by a notary and include such things as contracts, lawsuits, wills and other legal documents. While these tell us a great deal about the commercial and legal end of the business of whaling, most appear thus far to be remarkably silent on exactly how whales were hunted and processed in Terranova and are even less informative about what day-to-day life was like for some of the first Europeans to exploit the resources of what is now eastern Canada.

What the documentary sources lack, however, is at least partly preserved under the ground and beneath the sea at Red Bay. The shipwrecks, work buildings, living sites, cemetery and other areas have often produced eloquent testimony to the conditions of life and work in sixteenth century Butus. In this way the archival and archaeological sources compliment each other to reveal a much more nearly complete picture of Basque whaling and whalers than either source of information could hope to do by itself.

Archaeology at Red Bay

Archaeologists carefully remove the soil and refuse from around a shore station. Each find is recorded with reference to its horizontal position and the layer in which it was found.

After the initial survey in 1977, archaeological excavations began in a modest fashion in 1978. A crew of six from Memorial University worked on Saddle Island for a month that summer and in the early fall a crew of about the same size from Parks Canada began their underwater explorations. In the years which followed the scale of the work expanded dramatically; in the early 1980s as many as 60 people were employed on the land and underwater excavations. They included those doing the actual excavations, registering and conserving the finds, recording structures as revealed in the ground and underwater, drafting each ships' timber and recording the details of its manufacture, studying the barrel parts and other finds, analysing the whale bones, and a variety of other tasks including housing and feeding the crews and maintaining the equipment necessary to conduct major archaeological excavations.

Despite outward appearances, the basic methodologies of both land and underwater archaeology are remarkably similar. In essence, archaeology is a process of destruction. As soon as the first shovelful of earth is removed or the first layers of silt disturbed, evidence of human activity at archaeological sites is being destroyed and can never be replaced. For this reason excavations must proceed with extreme caution and every bit of evidence, no matter how insignificant it may seem at the time, must be carefully recorded.

Horizontal recording is done by establishing a grid of squares over the entire site—using wooden stakes on the land and a metal frame under water—within which the location of each artifact or bit of information can be plotted. Vertical recording is also important and depths below the ground surface and sea floor are carefully recorded. Even more important than depth, however, are the layers built up over time by natural processes or human activity. They not only provide evidence of chronology—the lower layers are older than those above—but may also reveal how a single area was used for different purposes at different times. What was once a campsite for prehistoric Indians or Eskimos may have been the location of a cooperage during the sixteenth century, a seasonal fishing station during the nineteenth century and a dump during the early years of the present century. The layering, or 'stratigraphy,' under water, on the other hand, revealed, in one case, an opposite situation. A single layer of codfish and whale bones indicated that whaling and codfishing were carried out simultaneously during the sixteenth century.

A diver 'fans' silt from a ceramic vessel. The air lift in the diver's left hand removes loosened silt in a fashion similar to that of a vacuum cleaner.

Denis Pagé photo courtesy of Parks Canada.

Aluminum frame divides the underwater site into one by two metre grid. As on land, objects are recorded with reference to the fixed grid and their depths and the layers in which they were found carefully noted.

Denis Pagé photo courtesy of Parks Canada.

The association of artifacts with one another is as important as their positions within the various layers which make up a site. Coopers' tools associated with broken cask parts, offcuts from making headpieces, shavings and even sawdust clearly indicate an area where barrel making took place. Underwater archaeologists found an object resembling a checkerboard, which was initially identified as a 'gaming board.' However, this gaming board proved to be carved on the ship's binnacle which held the navigational instruments. Its association with the compass, a sand glass and what appears to be a log reel caused archaeologists to reconsider their earlier guess. It now seems possible that the 'gaming board' may be some sort of navigational aid, perhaps a primitive counter or calculator whose purpose and instructions for use were forgotten sometime between the sixteenth and twentieth centuries.

In order to reveal information of this kind each layer of soil or silt must be carefully peeled away. On the land this is done by hand using masons' trowels and other small

A portion of the ship's hull was moulded 'as found' underwater. The mould was used to create this cast which shows the keelson (large timber near centre) and interior 'ceiling' planking.

Rock Chan photo courtesy of Parks Canada.

tools. Underwater, silt can sometimes be removed by 'fanning' with the diver's hand, but often trowels and other tools familiar to land archaeologists are also used. The only real difference, aside from the environment in which the archaeologists work, is in the way the soil is removed from the excavation area. Land archaeologists carry their backdirt to the spoil heap in buckets; the divers simply bring the mouth of the air lift near the area and the silt is removed by suction, a process very much envied by their colleagues on land.

Recording is essentially the same both on the land and under water. Each artifact is plotted within the grid. Unusual or important artifacts, all structures, ships' timbers and other large or immovable features are drawn and photographed in the positions in which they were found. In some instances at Red Bay very important features were molded in rubber, both on land and underwater, so that excavations could proceed beneath them. Later, casts were made from the molds which reproduced each particle of stone or wood, including tool marks made by the shipbuilders more than 400 years ago. Only after this careful recording has taken place can the objects and features be removed for conservation and restoration and to see what other information and objects lie buried beneath them.

Conservation: saving the finds

Virtually every artifact found at Red Bay requires specialized treatment if it is to survive for analysis and exhibition. This is so because the objects, whether on land or under water, have more or less reached a state of equilibrium with the environments in which they are buried. They may continue to deteriorate slowly in their burial environments, but the rate of deterioration is rapidly accelerated when the objects are removed from the ground or the water and placed in the open air. Some types of material are permanently damaged by only a few minutes exposure to air.

To make matters worse different materials buried under different conditions react in different ways to both their burial environments and to exposure to the air. At Red

Conservator Judith Logan of the Canadian Conservation Institute lifts an iron knife encased in wax from the refuse surrounding a cooperage. Hundreds of objects which would not otherwise survive are preserved and restored each year by conservators.

A human burial containing preserved wood and textile is removed from the ground in one piece by freezing with dry ice. 'Block lifting' complex features and artifacts allows them to be removed to the laboratory where careful examination and excavation under controlled conditions often reveal information which might have been lost by excavating them completely in the field.

Bay the types of materials found include virtually every substance utilized by sixteenth century Europeans. Stone, iron and other metals of all kinds, ceramics, glass, bone, wood, baleen, leather and textiles have been found preserved in almost every imaginable burial environment. The combinations of materials are almost endless; so are the possibilities for something to go wrong.

Fortunately for archaeologists there are specialists whose profession it is to preserve archaeological artifacts from the moment of discovery until they can be stored or exhibited safely under normal conditions. Conservators, as these specialists are called, from Parks Canada treated the material recovered from under water and the Canadian Conservation Institute provided conservation services for Memorial University archaeologists.

The conservator's job often begins the moment an artifact or other object is revealed by the excavators. In many instances objects are too large or too fragile, or both, to be simply removed from the ground and sent to the field laboratory. A number of techniques have been used at Red Bay to remove such objects. Small specimens, for example shattered pottery or badly deteriorated iron, are removed by coating the object and surrounding soil with a layer of wax or plaster, separated from the object by aluminum foil. The wax provides support so that the entire block can be removed. Some objects, for one reason or another cannot be treated in this fashion. In one case a suit of clothing consisting of a shirt or jacket and a pair of knee-length breeches was removed by covering the entire surface with small strips of gauze using an adhesive which could be easily removed later. The gauze provided support for the fragile textile and the entire costume could be removed as a single piece. In other cases entire human skeletons and their degraded clothing and other objects were frozen using dry ice and lifted into specially made padded containers.

At Red Bay the conservators' work includes maintaining the various excavated materials under conditions which will allow them to remain undamaged until they can be treated. As a general rule conditions as close as possible to those in which the objects were found are maintained. All metals and many other objects are routinely X-rayed in the field to determine what lies beneath the layers of corrosion and how severely the objects have deteriorated.

Some small artifacts can be stabilized in the field laboratory, but many others require specialized treatments in laboratories specifically designed for the purpose. The treatment of waterlogged wood, for example, involves replacing the water with a water-soluble wax and freeze-drying the artifact to remove the remaining water; the entire process can take several

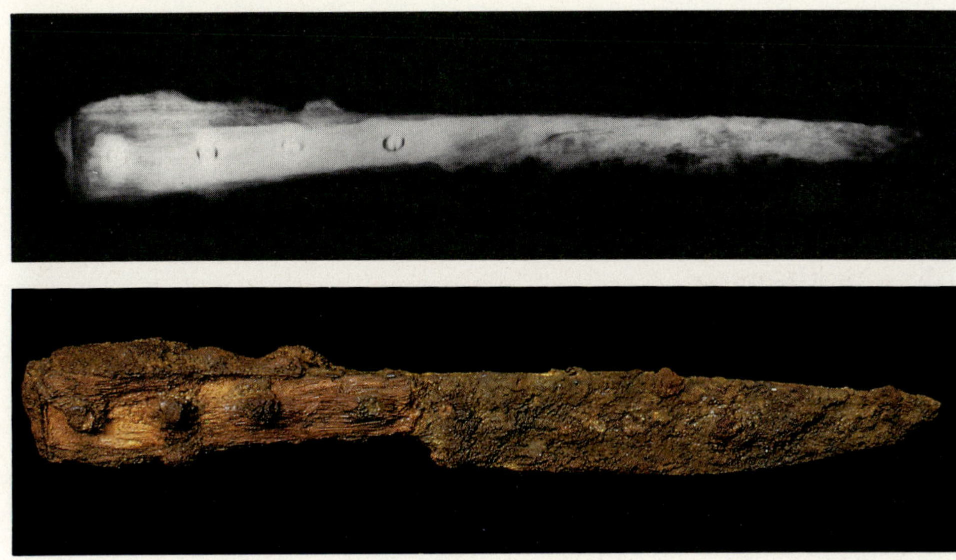

A seemingly well-preserved iron knife is revealed in an X-ray to consist largely of corrosion. Only the light portion of the blade contains metal. If the corrosion were removed, little would remain of the artifact. Metal artifacts from Red Bay are routinely radiographed to determine their form, condition and the most appropriate conservation treatment technique.

years. Metals are treated using a variety of techniques which involve removing the corrosion to the original surface layer and stabilizing the objects so that they may be examined and displayed. Even such seemingly stable artifacts as ceramics, glass and even stone are often so deteriorated that they require consolidation.

Once the artifacts are stable in normal conditions, many require restoration before their original forms are apparent. Ceramics and glass are prime examples of this type of artifact. Cooking pots, storage jars, pitchers, and drinking glasses are almost always found as small fragments. The small sherds which are all that remain of these objects must be carefully cleaned and separated into individual vessels. Next fragments which join to one another are identified and glued. When all the sherds which remain from a single vessel are assembled, it is unusual to discover that all, or even most, of the pot or bowl is

An example of the conservators'art, this small jar was reassembled from many tiny sherds. Missing portions were replaced with plaster of Paris painstakingly shaped to the vessel's original contours. Less than ten centimetres high, the jar probably held condiments or medicines.

present. In these cases conservators fill the gaps with plaster, or epoxy resins in the case of glass, and carefully shape the medium to the original form of the vessel. The surfaces of the filled portions are then painted to match the original surface, or at least coloured so that they are not obtrusive to the eye. It is not unusual for a single ceramic piece to require hundreds of hours of work; the restoration of glassware is often even more time-consuming.

As a result of the painstaking work of the archaeologists, conservators, and all of those other people connected with the Red Bay projects a great number of artifacts, and an even greater amount of information, has been recovered and preserved from the sites of the sixteenth century Basque whalers. The following sections of this book describe some of these finds and indicate some of the things archaeologists have learned from them.

Ships

Explorations beneath the frigid waters of Red Bay Harbour have revealed the wrecks of three whaleships, a mid-size sailing vessel, and several smaller boats. This is surely one of the largest collections of sixteenth century watercraft from any single location in the world; the cold waters of the Strait of Belle Isle and rapid deposition of silt over the wrecks have combined to make the ships and boats from Red Bay the best preserved wrecks of the period.

School of capelin flashes past a diver standing on the collapsed hull of the whaleship. Located a short distance off the shore of Saddle Island and in about ten metres of water, the ship is the best preserved example ever found of the 'workhorse' vessels which allowed European expansion to the New World.

Denis Pagé photo courtesy of Parks Canada.

A short distance off the shore of Saddle Island were found the remains of the first of the vessels located in Red Bay Harbour. After a remarkably short search in the fall of 1978 divers spotted the timbers of a wreck protruding through the silt of the harbour floor in less than ten metres of water. Preliminary excavations revealed that much of the hull remained in remarkably good condition and that the vessel appeared to be filled with collapsed barrels made from oak and beech and bound with wooden hoops. The casks matched exactly fragments of barrels recovered the preceding summer from good sixteenth century context at one of the nearby land sites. Even at this stage there was no doubt that the ship was one of those lost during the Basque whaling episode in southern Labrador. The fact that she was loaded with barrels, presumably once filled with whale oil, pointed to the possibility that the divers had actually located the *San Juan*, a ship laden with about 1,000 barrels of valu-

Trench through the whaleship reveals staves from collapsed barrels. The fact that the ship was fully loaded corresponds with testimony of whalers who witnessed the sinking of the *San Juan* in 1565.

Denis Pagé photo courtesy of Parks Canada.

able oil which was lost in a fall storm in the year 1565 as she was about to sail for Spain.

Beginning with those preliminary findings in the fall of 1978, divers from Parks Canada'a Marine Archaeology Unit spent eight years excavating the wreck of the *San Juan* and exploring the waters of Red Bay Harbour for additional evidence of the whalers' presence.

Surface support was provided by a barge moored over the wreck site. It was eventually equipped with a system for forcing hot water through the divers' suits to permit them to work in the cold water, which ranged from -2°C to about +5°C during the summer season. Silt which buried, and preserved, the wreck was removed with the aid of airlifts so that what remained of the ship and her cargo could be carefully mapped and photo-

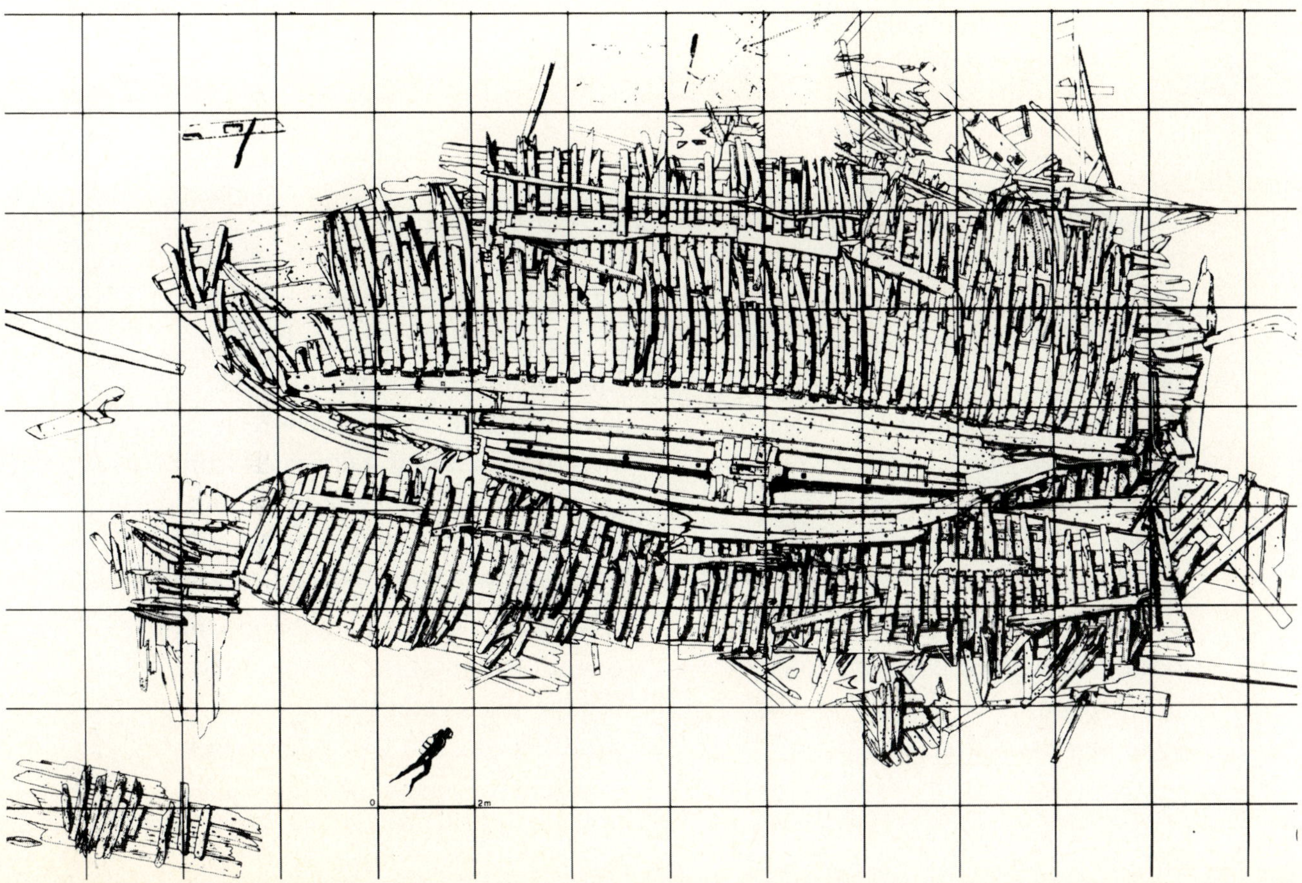

Plan of the excavated wreck almost resembles an architect's drawing. Clearly visible are the keelson and reinforced main mast step. To the right is the triangular transom with the corner of a single gun port. A carving of a ship found on the vessel also shows a single gunport, suggesting that it may be a representation of the *San Juan*.

Illustration courtesy of Peter Waddell, Parks Canada.

The ships's compass is carefully uncovered (left). Body and cover were turned from wood; the iron needle rotated on a brass pin set in lead beneath a glass cover. Brass gimbals kept the compass level despite the roll of the ship.

An artist's reconstruction of the compass is shown at the right.

Denis Pagé photo and Carol Piper drawing courtesy of Parks Canada.

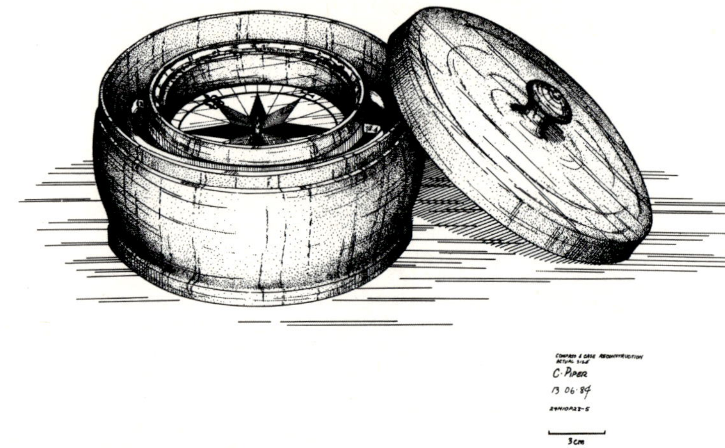

graphed prior to removal of artifacts and, eventually, the ship herself.

Since the upper works of the ship were above water for some time after she drove ashore many objects and parts of the ship itself were salvaged before the ship disappeared beneath the water and silt. Immediately after the storm, and during the following summer the ship's outfitter, Joanes de Portu, managed to remove most of the ship's gear and about one-half of her cargo. Some of the oak timbers used in the construction of buildings on shore may have come from the *San Juan*. Nevertheless, archaeologists recovered more than 200 barrels, or *barricas*, navigational instruments, some personal possessions and a large portion of the ship's hull, all in a remarkable state of preservation.

The ship itself had grounded stern first and eventually opened up lengthwise along the keel and stem. When the wreck was fully exposed it emerged from the fine silt of the harbour bottom almost as if laid out in a blueprint. The keel measured approximately 15 metres (50 ft.) in length and the overall length of the vessel, including the spars and rigging, was probably close to twice that figure. Its capacity was in the neigh-

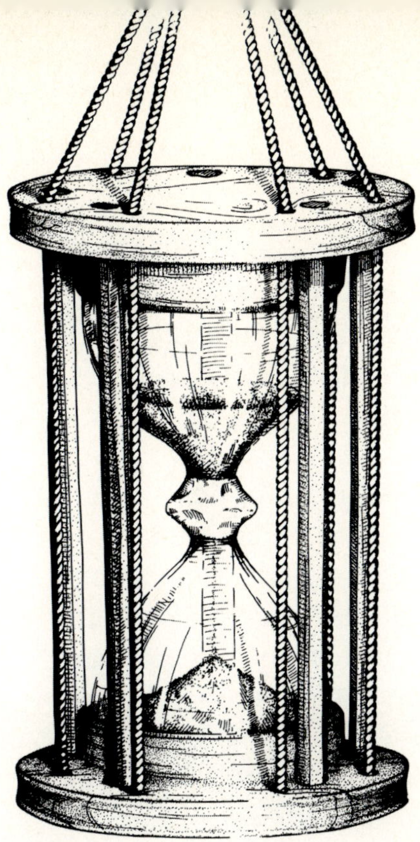

A sand glass is uncovered (right) in close association with the compass. It was used to determine the ship's speed by timing the rate at which a 'chip' attached to a log line was payed out over the stern of the moving vessel.

Artist's reconstruction at left.

Denis Pagé photo and Carol Piper drawing courtesy of Parks Canada.

bourhood of 250 tuns burden, which agrees with the recorded size of the *San Juan*. The ship was heavily framed of white oak and planked with the same material. She carried three masts and had a single gun port in her transom.

The latter features became important in the fall of 1983 when a diver recovered a softwood plank from the wreckage of the ship. On the plank was carved the profile of a three-masted ship with a single cannon protruding from the stern. It was scribed with a double-pointed instrument, possibly a navigator's dividers, and may be a representation of the *San Juan* herself.

Rather than attempting to raise the remaining portion of the hull as a single unit, Parks Canada archaeologists chose to dismantle the wreck and recover the timbers, planks, knees and so forth individually. By following this procedure a great deal was learned about the construction of the ship. A ship must be dismantled in something close to the reverse order to which it was built. By doing so archaeologists not only learned much about the ship's construction but also gained an appreciation of her builders' skill—some planks had to be removed with 12-ton hydraulic jacks.

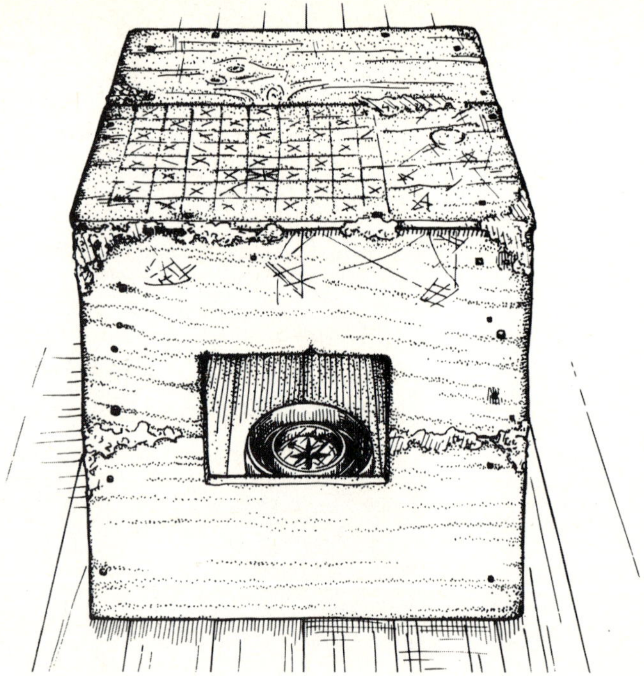

Artist's reconstruction of the ship's binnacle which held the compass and other navigational equipment. The checkerboard carving on the top was originally thought to be a gaming board, but may actually be be some form of reckoning device used in navigation.

Carol Piper drawing courtesy of Parks Canada.

Once the timbers were removed from the harbour a detailed drawing was made of each piece. The drawings recorded not only the dimensions and shape of every plank and timber but also the locations of nail and dowel, or 'treenail,' holes, tool marks, wear patterns and breakage. Following the exact recording of each part of the vessel, the timbers were returned to the bottom of the harbour where they remain today. The timbers were reburied under a thick layer of silt and sand; recent tests indicate that the same conditions which preserved the hull for more than 400 years now exist in the reburial area. Each timber is numbered and if technology and funding ever become available to conserve and restore the remaining parts of the vessel, the well-preserved timbers could easily be re-excavated.

The drawings of the timbers were then used as the basis for the construction of a model of the ship. Shipwright and model maker Marcel Gingras first modeled each part of the vessel at a scale of 1:10. Then, based upon the positions of the timbers on the wreck site, or by matching nail and treenail holes, the parts were assembled in their original positions. The result was a hull with surprisingly fine lines, one capable of carrying a heavy cargo but still making a good speed during the North Atlantic crossing. Although the ship lacked many of the refinements of larger warships of the time, no sacrifice was made in strength. Had it not been for the misfortune in Red Bay Harbour the ship would undoubtedly have made many more crossings from Spain to the New World.

Perhaps the greatest importance of the ship lies in the fact that it was not one of the elaborate warships often depicted in detail in sixteenth century drawings and paintings. Instead it is the best preserved example yet found of one of the 'workhorse' vessels which allowed the European expansion into the Americas during the sixteenth century.

Despite the salvage which took place shortly after the ship was wrecked, some objects remained buried in the silt of Red Bay Harbour until the twentieth century. Somewhat surprisingly, several navigational instruments escaped the efforts of the salvors. Near the ship's binnacle were found: the remains of a small compass, set in a turned wood case and covered by fine glass; the frame for a sandglass used to reckon time; and a wooden reel from which a line and 'chip' of wood may have been

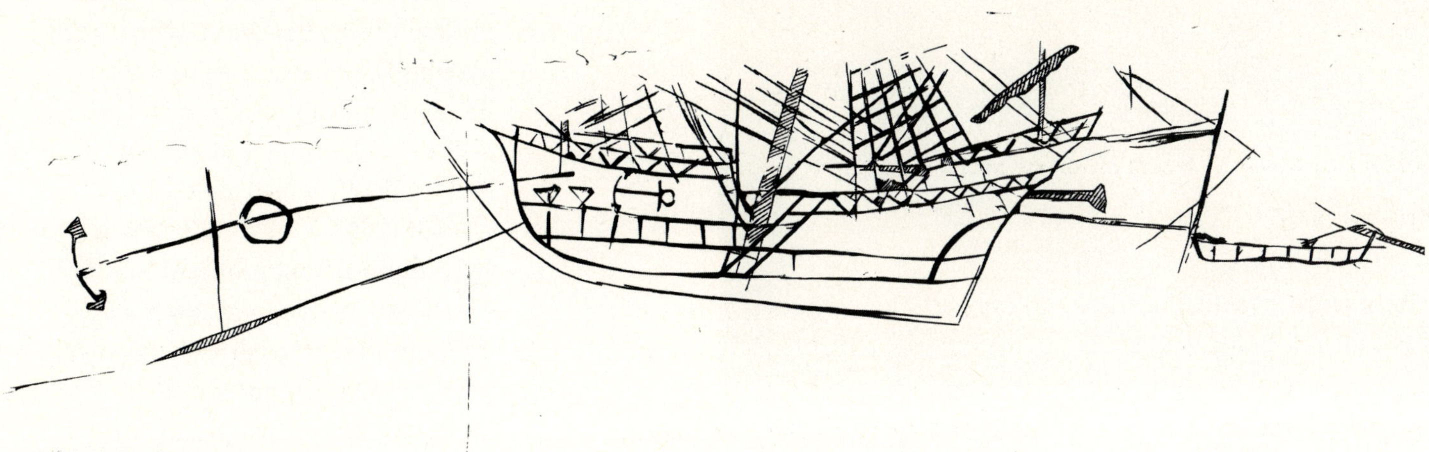

Carving on a softwood plank may be a representation of the ship revealed by Parks Canada divers. Anchor lines extend from the bow and a small boat is tied astern. The single gun protruding from the ship's stern matches the single gunport found on the transom of the Red Bay ship. Although partly destroyed by marine organisms, the carving also provides clues to the ship's rigging.

Carol Piper illustration courtesy of Parks Canada.

Although parts of the ship were above water for several years after it grounded, and many parts were salvaged, divers recovered many parts of the rigging and upper works. Here a diver displays a large block used to adjust the sails.

Denis Pagé photo courtesy of Parks Canada.

uncoiled to estimate the ship's speed. In a second wreck, on the opposite side of the harbour, was found a brass astrolabe, a forerunner of the sextant, used to determine a ship's position at sea.

Also found among the wreckage of the *San Juan* were rigging blocks, some with the rope still intact, a round oak timber, probably a mast or other spar, the ship's capstan and rudder.

A few of the crewmembers' personal possessions also were lost when the vessel sank. Ceramics include narrow-necked jars which may once have held wine or cider and a fine two-handled porringer with a blue pattern on a white background. The latter was found beneath the ship's sterncastle, the officers' quarters, and may have belonged to a man of substance.

No human bones were found in the wreckage, which accords well with the accounts of the sinking of the *San Juan*; all those aboard managed to reach shore safely despite the violent storm. There was a tragedy of a sort, however, for a wicker basket containing the bones of salt codfish and those of a black rat was recovered from the wreckage. Dr. Stephen

A piece of the puzzle is fitted to a model of the ship. Drawings of each timber were converted to 1:10 miniature ship parts. Based upon their positions on the sea floor, matching breaks and holes for fastenings, the pieces were reassembled to reveal a hull of remarkably fine proportions.

Denis Pagé photo courtesy of Parks Canada.

Cumbaa speculates that the rat may have been enjoying what proved to be a last meal of salt fish at the time the ship was driven aground.

At least two other sixteenth century ships lie on the bottom of Red Bay Harbour. Both that which produced the astrolabe and another, apparently somewhat larger, vessel are located in shallow water near the mainland shore opposite the major excavation site. These wrecks were only explored sufficiently to determine that many of the construction details duplicate those of the *San Juan* and that they are of the same period. They remain preserved in the silt and icy water as a storehouse of information for future explorations.

Excavations of the ship believed to be the *San Juan* concluded in 1985 when the last of the ship's timbers were reburied beneath the sea floor where they have lain for more than four centuries. The task of analysing the information and artifacts recovered from the wreck site continues, however, as it will for years to come. The result is sure to be a new look at the ships which opened up the New World to Europeans and at the men who built them.

Boats

Whaleships, or galleons, were not the only vessels discovered in Red Bay Harbour. Four other wrecks were also explored by Parks Canada archaeologists. One of these may have been a *pinaza*, or 'pinnace' in English, a common small sailing vessel of the sixteenth and seventeenth centuries. The example discovered in an inverted position off the west shore of Penny, or Organ's, Island in Red Bay Harbour was framed with oak and planked with elm.

The vessel, which measured more than ten metres in length and carried two masts, was probably used, among other things, for travel between whaling ports along the Labrador coast. The entire hull, up to and including the gunwales, was removed from the water and is being preserved by Parks Canada conservators.

Perhaps of even greater interest than the so-called *pinaza* are three small open boats, one of which may actually have been used in the whale hunt. Whaleboats were known as *chalupas* to the Spanish Basques, a word closely related to the English 'shallop.' The boats appear to have very fine lines and clearly were built for speed, an important attribute when approaching, and retreating from, a whale. The frames were sawed from natural curves of oak, and the boat was planked with light strakes of the same material, scarcely more than one-half inch thick. The upper strakes, above the waterline, were clinker-built (or lap-strake) while the lower portions of the small boats were carvel or edge-to-edge planked. It may be that this arrangement provided a minimal surface area below the waterline (for speed) and increased the surface area above the waterline for stability.

The small boat had positions for seven crew members, including the harpooner and steersman. There does not appear to be any evidence of a bollard or loggerhead, around which the harpoon line was put during a 'Nantucket sleighride,' in any of the whaleboats.

Like the *pinaza*, the small boats have been raised and are undergoing conservation. The finished products, and the information thus far recovered from the *chalupas,* promise to teach archaeologists a great deal about the origin and development of the boats used by more recent whalers throughout the world.

Nails and rivets long since dissolved, the frames and planking of a collapsed whaleboat rest on the bottom of Red Bay Harbour. Constructed from light timbers and plank, the boats were built for maximum speed and manoeuverability necessary in the whale hunt.

Denis Pagé photo courtesy of Parks Canada.

Shore Stations

Once the crews of the whaleboats had done their work, the whales were brought to one of the more than twenty shore stations which line the harbour-front. The preferred locations for shore stations were places which provided level land upon which to erect the tryworks, shelter from the northeast and southwest winds and deep water close to shore so that whales could be manoeuvered as near as possible to the scene of the actual rendering. Most of these locations are on the mainland along the eastern shore of Red Bay Harbour and on the southeast end of Saddle Island, facing the harbour. At the former location archaeologists have found remains of the stone tryworks literally stacked atop one another. When the earliest tryworks disintegrated to the point of uselessness they were simply levelled and a new structure built on the rubble. It is in this location

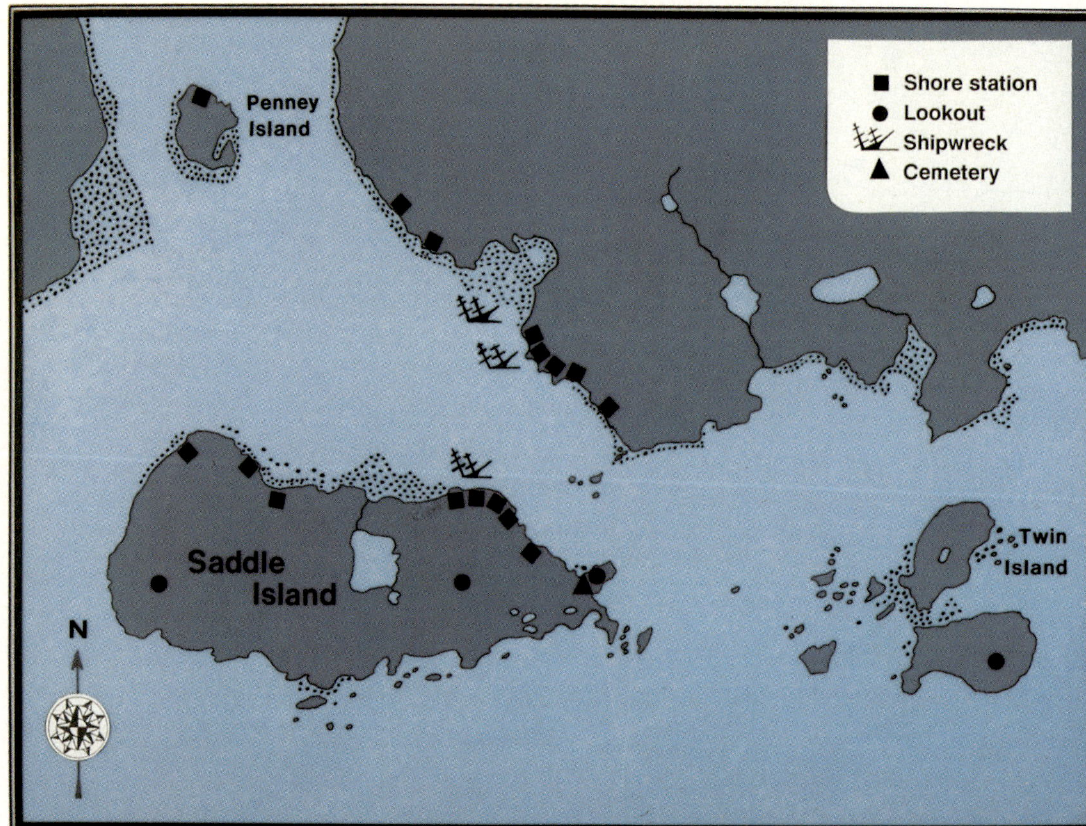

Map of Red Bay Harbour shows clearly the whalers' preference for locations sheltered from the northeast and southwest winds. Deep water close to the mainland shore (centre) provided an additional attraction; evidence of repeated rebuilding and intensive use has been found at this location. In contrast, shore stations in exposed areas (such as those at the far left) show little evidence of use. It is thought that they were built during the peak whaling years when more desirable locations were in use.

The lower stonework of the fireboxes is all that remains of this tryworks. Archaeologists have exposed the bases of the walls upon which large copper rendering cauldrons once rested. The fireboxes average about 1.3 metres in diameter.

that archaeologists expect to find both the oldest and most recent shore stations. Other locations were also utilized, such as the western end of Saddle Island and nearby Penney or Organ's Island. Since these often have the disadvantages of shallow water or lack of shelter, it is thought that they were used during the peak years of the whaling period—perhaps during the 1560s through the 1580s—when the more suitable areas were already occupied. In no case do these latter shore stations show the intensity of use that is obvious in more desirable locations.

The most conspicuous features of shore stations are the tryworks where the actual rendering was done. Often a low linear mound, parallel to the shoreline, marks the location of collapsed tryworks. Several of these have been left unexcavated and visitors to the site may see how the shore stations looked before investigations began in 1978. As the first sods are removed from one of these mounds blackened and fire-shattered rocks, often covered with burned whale fat, are immediately visible. Often they are covered by a thick layer of fragmented red tile—the remains of the collapsed roof of the structure.

Complete excavation of several tryworks has revealed a consistent pattern of their construction. The low rock mounds visible beneath the surface vegetation are, in fact, the collapsed walls of the firepits upon which the copper rendering cauldrons rested. The tryworks thus far exposed have fireboxes arranged in a linear fashion, always opening toward the beach, and numbering between one and six. The fireboxes themselves are between 1.2 and 1.5 metres in diameter and the entire stone structure may be as long as ten metres. Judging by the amount of burning in the individual fireboxes, it does not appear that all were in use at one time. Those at the ends often appear virtually unused, suggesting that more fireboxes were built than there were cauldrons. Then if in the course of the rendering process one of the fireboxes became too crumbled from heat to be usable, its cauldron could have been moved to a 'spare' firebox allowing the rendering process to continue almost without interruption. It appears as if, at least occasionally, the cauldrons might have burst during the rendering process for fragments of them have been found in the fireboxes. It is remarkable that neither documentary nor archaeological evidence suggests that the tryworks might occasionally have caught fire.

Fat-encrusted copper cauldron fragments were found in tryworks fireboxes. The black encrustation is on the outside of the cauldrons and resulted from using skin and fat which would yield no more oil as fuel for the rendering process. A rivet in the left-hand fragment gives a clue as to the construction of the cauldrons.

Some of the tryworks were dug into a sloping beach or terrace to provide a level working platform behind them. More commonly, however, the stone back wall was elevated above the surrounding surface and a wooden working platform extended back from the wall. Workers standing on this platform tended the cauldrons, adding blub-

Reconstruction of a tryworks at 1:10 scale shows the tile-roofed structure which sheltered the rendering operation. A single cauldron is shown as is the wood platform behind the fires where workers tended the cauldrons, adding minced blubber and removing oil as it was rendered.

ber as it was needed and removing oil and 'fritters,' as the completely exhausted tissue was known to later whalers, when it was necessary. Heavy deposits of hardened fat, which sometimes penetrated the ground to a considerable depth, attest to the fact that the rendering process was far from tidy. On one of the platforms were found collapsed tubs, made by sawing casks in half. In use they had been partially filled with water and the hot oil ladelled into them to cool and purify it.

It appears as if all of the tryworks were covered by substantial structures. In most instances a layer of roofing tile and scattered iron nails was all that remained of the structures. In one instance, however, a clearly defined set of post holes, which once contained posts up to 30cm in diameter, delimited one such structure. It measured more than eight by nine metres and, judging from the row of posts precisely in the centre, had a peaked roof. What provision for dissipating smoke might have been made is not known. There may simply have been smokeholes in the roof. If there were chimneys of some sort no evidence of them remains in the stonework of the tryworks; if present they probably began at the roofline.

It does appear as if the sides of the buildings were closed in. Although no trace of boards or planks remains, the archaeological deposits change abruptly from outside to inside the buildings at precisely the point where the walls should have been.

Archaeologists cannot be sure how many of the twenty or so known tryworks were in use at any one time; during the peak years it is possible that all were operating simultaneously throughout the summer and fall months. Although the smoke they produced must have offended both the eyes and nose, it signified success in the hunt and portended a profitable return to the Old World.

A strip of skin and blubber is lifted from a whale carcass at a wharf in front of a shore station. This reconstruction, based on information from excavations on Saddle Island, shows smoke billowing from the tryworks immediately behind the wharf and a small building which housed a cooperage on the hill to the right.

Painting by Richard Schlecht courtesy of National Geographic Society.

Coopers and Cooperages

Staves and heads of a barrel of 45 gallons capacity. Areas once covered by hoops appear lighter in colour than exposed portions of the cask.

Lester Ross photo courtesy of Parks Canada.

Coopers, or barrelmakers, were as important to the whaling industry as the harpooners or oil boilers. If casks were not tight the season's efforts would be lost as oil drained into the hold of the ship. Each whaleship had its cooper, or coopers, and assistants. They often began assembling casks from pre-fitted sets of staves on the crossing from Europe to Terranova. When the oak and beech staves were reassembled, head pieces were trimmed to fit the inside of the casks, grooves cut around the interiors of the ends of the cask, and the heads installed. Hoops were split from green alder, then notched and bound with withies to hold them together. Finally, the hoops were levered over the casks, where they covered about two-thirds of the outer surface, bung and vent holes were bored, and the casks were ready to receive their contents of whale oil.

Archaeologists have excavated two cooperages in their entireties and, as of this writing, have begun work at a third. All three cooperages are located on elevated terraces above, and where possible upwind from, the tryworks with which they were associated. Like the tryworks, the cooperages were also substantial structures with tile roofs. Within the cooperages, and particularly in the hillside dumps which surround them, archaeologists found a wide variety of coopers' tools and other objects.

Coopers' tools from Red Bay include those for all stages of cask assembly. Those shown here (from left) include a scriber with baleen handle used to mark assembled casks, two blades from a tool called 'croze' used to cut the groove into which heads were fitted, a head vise used to lift the head last piece into position before the staves were drawn tight and an adze used to hollow the insides of the ends of staves.

All phases of the assembly of barrels are represented. Iron wedges may have been used to split blanks for staves or headpieces from large billets of wood brought to Terranova for that purpose. The iron blades from large planes indicate that the edges of staves and head pieces were smoothed and fitted at the cooperages. This may have been done simply to replace a stave on a damaged barrel, but might also suggest the manufacture of barrels from wood brought from Europe. Adzes for hollowing and rounding the insides of staves have been found, as have drawknives used to smooth cask exteriors and to trim the rounded edges of headpieces. The blades from 'crozes,' the tools used to cut the grooves into which the heads are fitted, are among the most common artifacts. Small augers, or 'gimlets,' were used to bore vent holes and holes in the centre headpiece. Into the latter were threaded 'cask hooks,' or 'head vises' (see illustration) which allowed the last head piece to be lifted into place before the staves were winched tight. Wooden hoops, brought in bundles from Europe, were split and notched with knives then bound with withies and levered over the casks. Fragments of implements used to lever the hoops and to drive them into place have also been found.

Coopers not only worked, but also lived, in the tile-roofed cooperages. Small hearths, perhaps used to prepare food, are associated with some cooperages. All have produced vast quantities of ceramics used for food storage and preparation, drinking glass fragments, and other possessions of the men who worked and lived at them. The following section describes these and other finds related to daily life in sixteenth century Red Bay and presents the interpretations archaeologists have made from them.

A sixteenth century cooper at work. The casks depicted in this painting duplicate exactly those found preserved on the shipwreck and surrounding the cooperages on Saddle Island and the mainland.

Richard Schlecht painting courtesy of National Geographic Society.

Daily Life in Sixteenth Century Red Bay

One aspect of sixteenth century whaling in southern Labrador upon which the documentary sources are remarkably silent is day-to-day life. Lists of provisions for the voyages include such items as flour, peas and other foodstuffs and condiments, and ample supplies of cider and wine; except for the liquid refreshments the quantities reported often do not seem sufficient to have seen a whaling crew through an entire six month season on the Labrador coast. Other documents make occasional reference to the buildings which housed the tryworks and there are records of tiles being brought to repair the roofs of some buildings. Nowhere, however, is there mention of the buildings in which the crews lived, how they organized themselves along the shores of the Strait of Belle Isle, what they ate and how food was prepared, what they might have done for recreation, and so forth. Even to approach these questions we must depend exclusively on archaeological evidence—the bits and pieces of objects lost or discarded more than four hundred years ago.

During the early years of the investigations at Red Bay the absence of evidence for living sites suggested that the crews may have lived aboard the ships moored in the harbour during their entire stay in southern Labrador. Although archaeologists are now certain that this was not the case, it does seem reasonable to assume that watches were maintained aboard the ships to ensure their safety. Some of the ceramics found aboard the *San Juan* may be evidence of crew members or officers who spent the whaling season aboard ship.

Ceramics are among the most common evidence of everyday life at Red Bay. Cider and wine etched the lead glaze of the large pitcher (4) opposite. Also shown are storage jars (1, 3), a cooking pot (2) and two smaller pitchers. All were restored from many sherds found scattered at cooperages and other living sites.

Photo by Rock Chan.

Collapsed roof of poles (left), once covered with sods and baleen, sheltered workers at a shore station. In the foreground is a stave from a large cask or 'pipa.' An archaeologist cleans debris from the inverted 'floorboards' of a whaleboat upon which the harpooner stood.

Most of the evidence for daily life comes from scattered locations on the islands in Red Bay Harbour and on the mainland in the town itself. Generally speaking the pattern of settlement seems to have been a scattered one. In some cases dwellings appear to be associated with work areas, and we might assume that the men who lived in them were, for example, coopers or oil boilers. In other cases the living sites are scattered over the islands wherever a rock outcrop or other natural feature afforded some protection from the elements.

The clearest indications of domestic activity come from the cooperages where the barrelmakers worked. The floors of the buildings, and particularly the middens or dumps which surrounded them, are littered with broken ceramics, glass and the personal possessions discarded or lost by the men who lived there. Indeed, many of these objects—including those for food preparation—were probably personal possessions. At least no record of most such domestic objects appears in the provisioning lists for ships sailing on whaling voyages. The large numbers of artifacts found at the cooperages probably reflects to some extent the length of time such structures were in use. They were built using heavy posts and beams, had tile roofs, and were probably closed in on the sides; the impression is of more or less permanent buildings around which refuse might have accumulated for decades. While

Wooden tally stick was one of a pair used to record payment of a debt. The two symbols separated by a line are the personal marks of the individuals involved. Boxes in the matching grids on each of the pair of sticks were notched each time a payment was made. Cheating was impossible since identical notches were cut on each stick every time a payment was made.

Photo by Jack Martin, Memorial University of Newfoundland.

sheer numbers of artifacts may indicate only a long use of these structures, some evidence suggests that the coopers were somewhat better off than many of the other crew members. The sturdy nature of the buildings themselves must have made the coopers considerably more comfortable than did the huts or tents used by the whalers. Some of the objects found in the cooperages, for example white tin-glazed ceramics with blue, green or lustrous decorations, finely made glassware in Venetian style, coins and so forth, all suggest men of relatively high status.

Another living site associated with a work area consisted of a collapsed roof comprised of poles laid tightly together and covered with sods and strips of baleen held down by rocks. No tiles covered its roof and it does not appear as if the structure was anywhere near as substantial as the cooperages. It is located adjacent to one of the tryworks and may have been the residence of the men who reduced the blubber to marketable oil. Beneath the roof fall were found a few fragments of ordinary ceramic storage jars, the pits of olives and fragments of nuts, the 'floorboards' of a whaleboat, some split cane matting and various pieces of barrels and other fragments of wood. These few artifacts indicate an occupation of short duration, al-

Whale oil lamps of iron (left) and copper, both associated with tryworks, indicate that work went on day and night when whales were plentiful. The copper specimen has provision for four wicks and must have been used to light a large work area.

though the adjacent tryworks was rebuilt at least twice and must have been in use for several seasons.

The boat crews appear to have set up their accommodations some distance from the shore stations. The locations of these living areas were selected because the boat crews needed not only space for a dwelling of some sort, but also a sheltered beach at which their boat could be launched and landed and, if possible, some nearby hill which served as a lookout for whales. Suitable locations, either up or down the Strait of Belle Isle from the main port, may also have given boat crews living at them the opportunity to sight and pursue whales before the whales were seen by crews living closer to the shore stations. Archaeologists have little idea of what form the boat crews' dwellings might have taken, but once again they do not seem to have been very substantial. Scraps of wood and nails found at some locations suggest wood framed structures, perhaps covered with baleen or sail cloth.

For the most part the objects found within and around these dwellings consist of scraps of ordinary pottery, bits of discarded food bone, charcoal and other unremarkable detritus. In one instance, however, a small pond adjacent to one of these sites was used for waste disposal. In the pond were found a number of wooden artifacts not ordinarily preserved at sixteenth century sites. These included spits, probably used to roast meat, which were made from barrel hoops and other woods, chopped and burned whale bone which served as fuel, fragments of wood from ships and small boats, a number of poles which may have been from the

Although documents indicate frequent financial transactions among the whalers, coins from Red Bay are extremely rare. The three examples shown above (from left) were struck during the reigns of Ferdinand and Isabella, probably in the early 1500s, Philip II of Spain (1556-1598) and Henry IV of France (1589-1610).

Sport and sustenance were derived from this object resembling a table tennis paddle. As shown in the photograph it is still used in the Basque country. Thrown over the heads of flocks of doves, it frightens them into nets strung below. It was found with the bones of dovekies, native Labrador birds which may have been hunted in the same fashion.

Photo courtesy of W. Douglas, Basque Studies Center, Reno, Nevada.

structure itself, fragments of a remarkably fine drinking glass, and a number of other objects.

The most numerous living sites found at Red Bay are also by far the least substantial. All take advantage of some natural feature of the topography. Some incorporate a near-vertical bedrock outcrop as a wall of some otherwise impermanent structure, but most seem to have been built around small cracks or 'corners' in the bedrock which served as natural fireplaces. All that remains are bits of wood charcoal and burned whale fat, a few nails, scraps of baleen and occasional fragments of ceramics and bits of glass. The nails suggest wood framed structures, perhaps once again covered by baleen or cloth. It hardly seems likely that these flimsy structures were occupied during the entire whaling season; perhaps their occupants retreated to more substantial structures, or to the ships themselves, during the cold fall and early winter months.

Although some members of the whaling crews appear to have enjoyed more comfortable accommodations than others, many aspects of their lives during the whaling season may have been very similar to one another. While the archaeological evidence is far from perfect as relates to the matters of diet and so forth, a few scraps of evidence allow us some guesses at such matters.

From the available evidence it appears that most members of the whaling crews were left largely to their own devices insofar as food, as well as shelter, was concerned. Except for cider and wine, the ship's provisions often do not seem to have been sufficient to provide for the entire crew for up to six months, and it seems unlikely that

Sherds of glass, many less than 0.5mm thick were reassembled into this fine drinking glass. It stands 15cm high, has a capacity of about 13 oz. and weighs less than a hen's egg. Its place of manufacture is not certain, but it is in the Venetian style. Without doubt it was a prized possession of the whaler who brought it from the Old World to the New.

Personal possessions lost or discarded at Red Bay include a tin-glazed porringer and wood comb, both from the wreck site, three lead tokens inscribed with personal marks, a copper finger ring and a stoneware cup or mug from northwestern France.

Porringer and comb photos by Rock Chan courtesy of Parks Canada.

each crew member could have brought more than a few 'luxury' food items with him. Rather, the whaling crews seem to have subsisted to a large extent on what was available in southern Labrador.

Archaeological evidence suggests that this included codfish, capelin and probably salmon, birds of every kind from dovekies to the now-extinct great auk, and probably some mammals as well. Seal bones are associated with some sites, polar bear remains have also been found and it seems very unlikely that the tonnes of whale meat discarded as a by-product of the oil industry would have been ignored. A typical meal might have consisted of little more than meat, fish, or fowl roasted or prepared in a stew seasoned with personal supplies of condiments, and perhaps supplemented by ships biscuit or bread for as long as it lasted. Without doubt the food was washed down with cider, and perhaps wine in the case of some individuals; the ships stores contained barrels of these liquids and many whalers may also have brought personal supplies. In season the wide variety of berries which still grow on the whaling stations almost certainly added variety to the meals.

Both the practice of leaving the crewmembers to their own devices to obtain most of their food and the resulting diet itself, may seem unusual to us. In fact, the protein-rich diet was probably enjoyed by the whalers, whose relatives in Europe subsisted on considerably less meat and fowl; the abundance of fish and game, as

well as the meat of whales, probably ensured that hunger was not a problem. Such a protein-rich diet can have its disadvantages, however, and these became apparent during some years when the rapid onset of winter trapped ships in the ice and their crews were forced to remain in southern Labrador. Deaths were common in those years, but many of them occurred in the following spring or even early summer, the result not of starvation or freezing but of scurvy, against which the high-protein diet did not protect the overwinterers.

Not all the time in southern Labrador can have been spent in the pursuit of whales and related activities. Although no 'recreational' artifacts (except for wine and cider glasses) have been found at Red Bay we can be sure that they were present. Games of chance, popular in the Basque country to this day, must have been played during idle hours. The harbour must have heard the sound of the traditional Basque three-holed pipe, the *txistu*, and drum, perhaps to celebrate the first whale, a full cargo, or some other occasion.

Since at least the time of Hannibal the Basques have been known as dancers, and the recreational and ritual dances of their homeland, often performed solely by men, were undoubtedly a part of life at the whaling stations. Although far from home, or perhaps because of the distance from Euskal, traditional Catholic festivals and feasts, often with ancient pre-Christian overtones, undoubtedly were celebrated in Labrador with the same fervour as they are in the Basque country today.

Death on the Labrador coast

'...I, Juan Martinez de Larrume...in bed and ill...ordain and establish this my testament...' So reads the will of Juan Martinez done in Butus on June 22, 1577, the earliest surviving will written in what is now Canada. The previous winter set in so rapidly that many ships were trapped in the ice and forced to overwinter. Many deaths followed, including that depicted to the left which shows a whaler dying of scurvy dictating his last will and testament. In the background a burial party attends to its solemn duties on the southern end of Saddle Island.

Richard Schlecht painting courtesy of National Geographic Society.

Whalers' cemetery on Saddle Island restored to its sixteenth century appearance. The rocks which marked and covered the graves were replaced and left exposed after archaeologists completed their excavations. About one-fourth of the more than 140 skeletons were removed for study; the remainder were reburied and still lie where they were first interred more than 400 years ago.

Despite the financial gains to be had from whaling, and the respect whalers enjoyed in their homeland, the business was a dangerous one. The North Atlantic crossing itself could be an ordeal. Provisions for creature comforts aboard ship were minimal, especially on the return voyage, often during the winter months, when the valuable cargo of oil had priority over the men who had obtained it. Life at a whaling station must have been equally demanding; hours were probably long and wind, rain, snow and blackflies often made leisure hours uncomfortable. Worse than mere discomfort, however, was the ever-present possibility of serious accidents or disasters such as the forced overwinterings. Death must not have been uncommon even among crews chosen for their skill and stamina.

In 1982 archaeologists located a whalers' cemetery on the extreme southern end of Saddle Island. During the following four years some 62 graves were excavated containing the skeletons of more than 140 whalers. Although

A well-preserved burial is exposed. About one-half of the graves contained a single skeleton; the remainder between two and a dozen. With the exception of two children aged about 12 years all of the skeletons appear to be those of adult European males.

some graves contained more than one skeleton, the number of burials found averages only about one per year for the period 1540-1600. Unless the death rate was remarkably low, or there were other means of disposing of the dead, there must be other cemeteries around Red Bay that archaeologists have been unable to find.

The cemetery is located adjacent to a small sheltered cove, once the home of a boat crew, in an area bounded on all sides by bedrock outcrops. For the most part the skeletons were found on their backs, heads carefully placed to the west, and with their hands crossed near the waist. Often several large rocks were found directly on top of the skeletons, perhaps placed there as grave markers, but also to prevent scavengers from disturbing the shallow graves, which averaged little more than a foot deep. Originally the rocks must have been visible above the ground surface, but the slumping of the grave fill and build-up of soil made all but a few invisible when the cemetery was found.

About one-half of the graves contained but a single skeleton; the remainder contained between two and a dozen individuals. With two exceptions, one a child about 12 years of age and another skeleton

Common grave of seven whalers may be the resting place of a boat crew killed during the hunt. At least one of the small boats found by Parks Canada archaeologists had positions for seven men.

of about the same stature, all the burials seem to be those of adult European males. Brenda Kennedy, the physical anthropologist who studied the skeletons, reports that they represented individuals who were short in stature but robust and well-muscled—precisely the sort of men who would have been selected for the rigourous and dangerous business of whaling.

In no case did the skeletons reveal cause of death, but this is a situation not uncommon in ancient skeletal populations. Circumstantial evidence, on the other hand, offers some suggestions. One large oval grave pit contained the remains of 12 or 13 whalers, who appeared simply to have been tossed into the pit with little regard for the careful arrangement of the skeletons which characterized most of the other graves. The best interpretation of this unusual feature is that the skeletons are those of whalers who died during one of the overwintering disasters, perhaps that recorded for 1576-1577, during which time the snow and frozen ground made burial impossible. No doubt the dead were buried the following spring when

A wooden cross was preserved on the chest of an almost completely degraded skeleton. It was found in a deep grave located outside the main cemetery; the cross is the only grave offering found in more than 60 graves. The reason for its inclusion still remains a mystery.

the ground thawed, either by survivors of the overwintering or by recently-arrived whalers from the Basque country.

Without doubt some whalers fell victim to whales they were attempting to kill. One grave contained a row of seven skeletons, clearly buried at the same time. Since at least one of the small boats recovered by Parks Canada archaeologists had provision for seven crew members, it is not impossible that this grave is that of an entire boat crew killed in an accident involving their intended victim.

Other graves probably also contain the victims of similar accidents and the skeletons of whalers who died from disease. One skeleton may even be that of Juan Martinez de Larrume who dictated his Last Will and Testament at Red Bay on June 22, 1577, the first day of summer following the disastrous overwintering of that year.

Additional evidence for an overwintering disaster comes from the first skeletons found in the cemetery. Carefully laid out in two rows were the remains of twelve individuals. They clearly had not been buried, but rested on the earthen floor of a building which once housed a boat crew. It appears as if the bodies were placed there during the winter, probably with the intention of burying them the following spring. Strangely, however, burial never occurred and the skeletons remained where they had placed to be enveloped by the slow-growing vegetation of Saddle Island. Why this was allowed to happen remains a mystery. Perhaps the event occurred near the end of the whaling period and no survivors remained to bury the dead nor did other whaleships arrive from the Basque country the following spring.

Preservation of the skeletons in the cemetery was extremely variable. In one area an ancient beach, formed long ago when the sea level was somewhat higher, contained particles of shell which preserved many skeletons virtually intact. In other areas, where the soil was both wet and acidic the bones had almost completely dissolved. The well-preserved skeletons, from which something could be learned and which numbered about 30, were removed; the remainder were exposed while physical anthropologists made their observations, and were then reburied beneath a layer of sifted shell and earth.

Shirt and trousers were preserved by the same conditions that dissolved the bone in this burial. Heavy wool trousers gathered at the knee were originally dyed dark blue. The shirt, with sleeves and collar still visible, was dyed red. The possession of dyed clothing indicates a man of relatively high status.

Jeremy Powell photo courtesy of the Canadian Conservation Institute.

In some cases the same wet and acidic conditions which dissolved the human bones acted to preserve other objects. Many skeletons, including those of the 'unburied dead,' were surrounded by scraps of textile. In one case a pair of knee-breeches and an upper garment, probably a shirt, were found in a shallow grave from which the bone had all but disappeared. Several skeletons were accompanied by thick round or oval lead discs averaging about 5cm in diameter and perforated by many small holes. They were found near the knees of the skeletons, with never more than one to a skeleton. Although it is suspected that they may have been sewn into the hem of a coat or cloak to prevent its blowing in the wind, the function of these artifacts remains unknown.

In 1986 explorations outside of the cemetery revealed five additional burials, three of them different in one way or another from the graves within the cemetery. In one case the deeply-buried skeleton had on its chest a large wooden cross, about 40cm high.

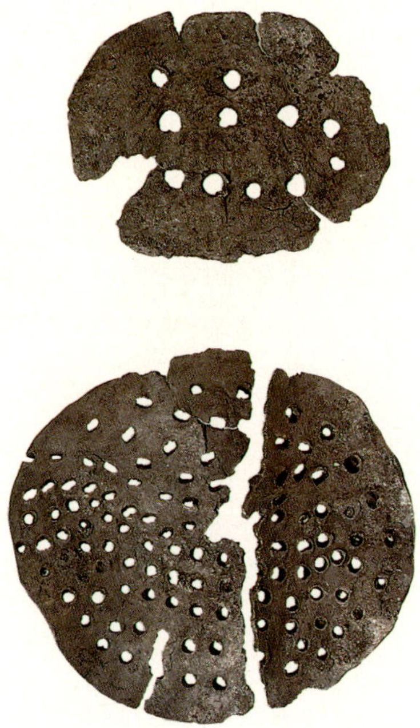

Lead discs of uncertain function were found with several burials. Always found singly and located near the knees, they may have been sewn to the edge of a cloak to prevent its blowing in the wind.

Photo by Jack Martin, Memorial University of Newfoundland.

Another burial, this one containing three individuals, was also unusual. The skeleton in the centre headed to the east rather than the west and, as excavation proceeded, many lumps of badly corroded iron appeared mixed with degraded textile and leather. At this point the entire burial was removed in a block (by freezing with dry ice) and sent to the Canadian Conservation Institute in Ottawa for further examination. X-rays showed that three of the iron objects were large keys and others long, slender, pointed objects which resemble the blades of daggers. As of this writing they have not been removed from the burial matrix; their condition is such that to do so might destroy them.

Still another 'burial,' if it can be called that for no grave pit was observed during its excavation, appeared to have been sunk in a shallow depression in a peat bog and weighted down with several large rocks. The acid soil of the peat bog had completely dissolved the skeleton but had preserved a set of woolen clothing—consisting of a knitted cap, a shirt and jacket, knee-length breeches, tailored stockings and leather shoes—in addition human hair, a beard and finger- and toenails. On the left side of the costume, unfortunately damaged by the rocks placed over the grave, was found a small, crudely-made iron thrusting weapon not unlike a dagger.

The wooden cross was the first grave offering, in more than 60 burials, which had been deliberately placed with one of the skeletons. The keys and 'daggers,' if not grave offerings, were at least not removed from the bodies prior to burial as must have been done with the personal possessions of the individuals within the cemetery. Why these objects were placed with the dead is not clear and the mystery is compounded by the fact that these burials were made outside of the well-defined cemetery. If the cemetery consisted of consecrated ground, which is likely since priests accompanied at least some of the whaling ships, there must be some reason why these individuals were not buried within the cemetery. They may not have been Christians, or perhaps not Catholics since there were heresies in the Basque country in the sixteenth century, or perhaps some more sinister factors were at work. It is not beyond the realm of possibility that the burials made outside of the cemetery were those of men who died in an unnatural fashion or had somehow violated the laws of the sixteenth century church.

Remarkably well preserved costume from a burial in a small bog includes a knitted cap, wool shirt, jacket, breeches and tailored wool stockings. The shirt and jacket were sewn from the same fabric into which was woven a plaid pattern of brown and white using natural wools from different varieties of sheep.

W. Bokman photo courtesy of the Canadian Conservation Insitiute.

Epilogue: Red Bay after the Basques

Fishing premises dating from the mid-nineteenth century still stands on Penney or Organ's Island. Both names derive from operators of a business which, during its peak years, shipped more than one million pounds of salt fish to many parts of the world.

Photo by Judith Logan.

Although the whaling period came to a close in the early seventeenth century, a harbour as protected as that at Red Bay must have continued to be used throughout the seventeenth and eighteenth centuries. In the former century it must have been used by codfishermen from many nations.

In 1715 a Quebec merchant, Pierre Constantin, established a *habitation*, probably a fortified trading post, somewhere around the harbour. The place was burned by Inuit in 1718, with the loss of three of its inhabitants. It was rebuilt in 1719 and apparently continued in use until at least the 1730s. In 1987 archaeologists found the remains of a large wooden structure and collapsed stone fireplace on the west side of the harbour; the building showed evidence of having been burned and the few artifacts found pointed to an early eighteenth century date. Absent, however, were glass beads and other items usually associated with fur trade posts, so it was not certain whether Pierre Constantin's *habitation* had been discovered. In 1988 archaeologists partially exposed a second chimney fall and associated structure only a short distance from the first, which also showed evidence of burning. At this site, however, there was found ample evidence of the fur trade: hundreds of pieces of lead shot of all sizes and tiny beads of blue and white glass can hardly have been intended for anything but exchange with Indians and Inuit for furs. Additional work is planned for the location to determine which of the two sites is that of the Quebec merchant's original post and which represents the rebuilding after the attack of 1718.

Flag of the Red Bay Cooperative Society founded by Wilfred Grenfell first flew in 1896. One of the first co-operative ventures in Canada, it survives today as the Red Bay Stores.

By the early nineteenth century seasonal fishermen were summering each year at Red Bay where they caught and salted codfish. They came from many places, but most were Newfoundlanders from Conception Bay. In many instances the remains of their dwellings and other buildings overlay the sixteenth century whaling stations. This material was carefully recorded and collected and comprises a remarkable collection of artifacts and information pertaining to life at Red Bay during the first half of the nineteenth century.

Eventually some of the Conception Bay fishermen and their families began to overwinter at Red Bay and the present community was founded, probably about 150 years ago. Many foundations from the early periods of settlement are to be found along the west side of The Basin and even some distance back from the shore where winter quarters must have stood. This aspect of the history of Red Bay has not been investigated archaeologically, but the sites exist and would form the basis for a fascinating study of the organization and re-organization of a single community over a century or more.

In the 1840s a large premises for the production of salt codfish was constructed on Penny Island in Red Bay Harbour. The island takes its official name from the Carbonear merchant family which first constructed the buildings. The island is today known as 'Organ's Island' after the present owners, who acquired the property early in this century and still maintain the buildings, some of which date to the time of the Penneys.

Fishing first attracted Europeans to Red Bay and it has remained the mainstay of the community until the present day. In 1896 Sir Wilfred Grenfell founded the Red Bay Cooperative Society, the first such venture in the province. The original flag now hangs in the Red Bay Store, successor to the original cooperative society. A new fishermen's cooperative now exists and the processing of many species in addition to cod promises to continue to provide a livlihood from the seas' resources as has been the case for more than four centuries.

Salt cod are still sun dried as they have been for more than four centuries, as shown in this photograph by Basque photographer Xabi Otero. Although the Basques, Red Bay's first fishermen, have long since deserted their ancient New World ports, resources of the sea continue to be a mainstay for residents of the Labrador coast.

Photo by Xabi Otero.

Acknowledgements

This booklet summarizes the labours and investments of a number of individuals and institutions. Selma Barkham first led Walter Kenyon, Ralph Pastore, Gerald Penney and James Tuck to the Basque whaling sites in 1977. The following year she encouraged Robert Grenier and his team from Parks Canada to search for Basque shipwrecks in Red Bay Harbour. Since that time literally hundreds of individuals have participated in the recovery and analysis of objects and information recovered from beneath the soil and waters of Red Bay.

Underwater explorations were conducted by Parks Canada's Marine Archaeology Unit. Excavations on the land were conducted by Memorial University with financial assistance from the Department of Culture, Recreation and Youth, Government of Newfoundland and Labrador, the Social Sciences and Humanities Research Council of Canada, the Department of Regional Economic Expansion, Government of Canada and the National Geographic Society.

The Canadian Conservation Institute has been a partner in the land excavations since 1978 and has provided expertise and assistance in analysing, preserving and restoring the thousands of finds; Parks Canada's Conservation Division did likewise for the material recovered from underwater. Without the assistance of these two groups the excavations would have been impossible. Staff of the Zooarchaeological Identification Centre, National Museum of Natural Sciences, analysed the whale remains and other bones from the various sites.

Many others assisted in one way or another with the excavation, preservation and analysis of the thousands of objects and bits of information recovered during more than a decade of excavations. To all of these individuals and organizations we extend our heartfelt thanks.

Finally, gratitude should be expressed to the many workers, far too numerous to name, who laboured to recover the information described briefly on the preceding pages and to the residents of Red Bay for untold assistance and, particularly, for never failing to make us welcome in their community.